6.0 Process Control in Speed Frame

6.1 Significance of Speed frame:

Speed frame process normally comes after comber in combed process and comes after draw frame in carded process. Speed frame is also called as simplex or roving frame. The speed frame process minimizes the sliver weight to a suitable size for spinning into yarn and inserting twist, which maintains the integrity of the draft strands. It is impossible to feed the sliver to ring frame for yarn production due to limitation in draft in ring frame. The irregularity of the output strand increases as the draft increases due to the noticeable effect of the drafting wave. Due to the above reason, the draw frame sliver hank must be reduced in two steps, so that an acceptable yarn quality is achieved. Cans of slivers from finisher drawing or combing are placed in the creel, and individual slivers are fed through two sets of rollers, the second of which rotates faster, thus reducing the size of the sliver. Twist is imparted to the fibres by passing the bundle of fibres through a roving "flyer". The product is now called "roving", which is packaged on a bobbin. A roving is a long and narrow bundle of fibrous strand. Roving is an intermediate product produced from sliver and it is normally used as a precursor for yarn. Roving is also distinguished as the first process in which material is wound on a bobbin. Faulty roving preparation has a drastic effect on spinning performance. The process parameters adopted in roving process has a significant influence on spinning quality and production. The speed frame machine essentially comprises between 60 to 132 spindles, each containing a drafting system and flyer twister. The rotation of flyer imparts twist to the fibrous strands. Flyer rotational speed is limited because of mechanical design difficulties. The defect arises in the drafting of roving introduces the short term irregularity in the yarn produced from it. The wrong selection of twist in the roving affects the spinning performance by either higher creel breakages or higher undrafted ends. The improperly built bobbin in roving leads to end breakage in ring frame and higher slough-off during material handling. The preparation of roving bobbin for the yarn spinning is of paramount importance for a spinning mill.

6.2 Tasks of Speed frame:

The tasks of a speed frame machine may be divided into number of individual operations, each of which is almost independent of the others. The major tasks of speed frame process are listed as:

- **Drafting**: to reduce the size of the strand
- **Twisting**: to impart necessary strength
- **Laying**: to put the coils on the bobbins
- **Winding**: to wind successive layers on the bobbin at the proper rate of speed
- **Building**: to shorten successive layers to make conical ends on the package of roving.

In a sliver of 3 ktex, approximately 20,000 fibres are present in its cross section. The draft of 10 would reduce the number of fibers to 2000 in the cross section, and a small amount of twist would be required to provide sufficient cohesion for suitable handling. Drafting is normally carried out by a draft system with double apron capable of working with entering sliver counts of 0.12 Ne to 0.24 Ne and counts of the delivered roving of 0.27 Ne to 3 Ne. Draft given in roving process is normally calculated from the hank sizes involved. The draft given in the roving process will be in the range of between 4 and 20 and can work fibres of a length up to 60 mm.

In addition to accomplishing drafting, the operation inserts a slight amount of twist to give the roving the required strength and puts the strand in a special type of package to facilitate handling. The insertion of twist in the roving must not create any difficulty in the ring frame drafting operation by developing a high drafting force. The roving with higher twist requires higher draft in the ring frame. The increase of draft beyond the recommendations deteriorates the yarn quality.

Lay refers to the arrangement of the roving coils wound around the bobbin in any given layer. The closeness of the lay is measured in "coils per inch" which means number of roving coils wound around the bobbin per inch parallel to the axis of the bobbin. The purpose of the laying operation is to put the successive coils of roving side by side in a uniformly spaced arrangement. This regular, uniform arrangement is achieved by the making the bobbins to move up and down at a uniform rate of speed for each layer.

Winding is the process by means of which the roving is drawn from the front roll through the flyer and onto the bobbin. The rate of winding compared to the rate of delivery at the front roll controls the winding tension. The building motion is controlled by the steady upward and downward movements of the bobbin rail containing the bobbins and spindles.

6.3 Importance of Machine Components in Speed frame:

6.3.1 Creel Zone:

Creel is the place situated at the back of the machine where the raw material placed for feeding to the drafting zone. Cans from comber or drawing machine are kept at the creel in an orderly fashion to utilize the floor space effectively. The position of the can in the creel corresponding to its drafting head is crucial in controlling the false drafts. The method of feeding slivers in speed frames and condition of cans are important. Slivers crossing each other, damaged edges of drawing sliver cans etc. would disturb the free withdrawal of slivers from cans. Sliver stretch in the creel in speed frames due to too higher creel draft has to be avoided. Optimum creel tension draft should be selected to control sagging or stretch in drawing sliver. Creel normally consists of 4 to 6 rows of guide rollers fitted with smooth plastic sliver guides running along the length of the speed frame. The draw frame sliver cans are arranged in 4 or 6 rows in the creel zone. The position of the plastic sliver guides in the guide rod corresponding to the drafting head is important in the control of stretch. The sliver guides to be checked during machine maintenance to ensure the production of uniform roving. The speed of the creel guide roller is crucial for combed sliver as it lacks cohesion. The wrong selection of creel tension draft for combed sliver may leads to more creel breakages. In modern fly frames, the creel transport rollers are arranged without vertical supporting rods and these types of creels are called telescopic creels. This type of arrangement enables placement of cans without any hindrance and also the movement of the machine operators is easy.

If a speed frame has 120 spindles, the cans fed at the creel may be in batches of 30 with different sliver content in the can. This makes the sliver can replenishment time as minimum as possible that does not affect productivity much. Sliver distribution from a single can during severe back material shortage is to be avoided to ensure better quality roving. Topping cans with the last few layers of the previous cans leads to production loss. When the slivers are spliced, the mass is usually not acceptable leads to a quality stop. The malfunction of any one creel guide roller may affect the roving quality in terms of false drafts. The correct functioning of all creel guide rollers should be checked periodically. The drive systems of creel rollers should be maintained properly during cleaning.

6.3.2 Drafting System:

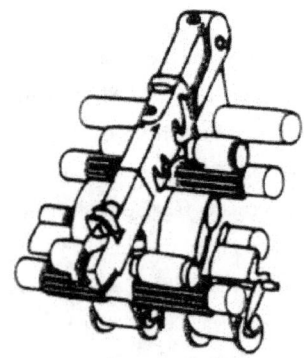

The purpose of drafting is to attenuate or reducing the weight per unit length of the feed material to the required fineness or count. The drafting system in the speed frame is usually consists of top rollers, bottom rollers, top arm and the associated parts. The important parts of drafting system of speed frame are given below:

- Bottom rollers
- Rubber covered top rollers
- Aprons (Top and bottom)
- Spacer

- Condenser (Inlet and Floating)
- Top arm

Shankaranarayana reported that eccentric drafting rollers, hard cots, lower pressure on top rollers and high roving irregularity increase the thick and thin places in the yarn.

6.3.2.1 Bottom Rollers

Bottom rollers have opposite helix flutes for zero axial thrust. The exact circularity of bottom rollers and top rollers prevents roving breakages and it can be duly interchanged. The rollers are completely hard chrome plated that helps to reduce the lapping. The bottom rollers of speed frame exercise immense influence on the quality of the roving. Eccentric or damaged bottom rollers, especially front bottom rollers, are the most common cause of unnecessary roving faults and excessive roller laps. Precise concentricity of bottom roller is a function of improved roving quality (evenness and strength). By means of electronical straightening, bottom rollers reach maximum concentricity. The running conditions of top rollers and bottom rollers are equally influential to roving breakages etc. Bottom rollers having narrow tolerances of dimensions B, p and T, produces fault free roving.

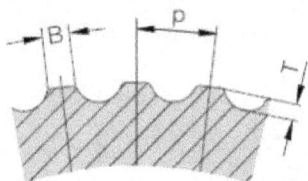

The spectrogram of yarn evenness produced from faulty bottom roller and fault free bottom roller are given below:

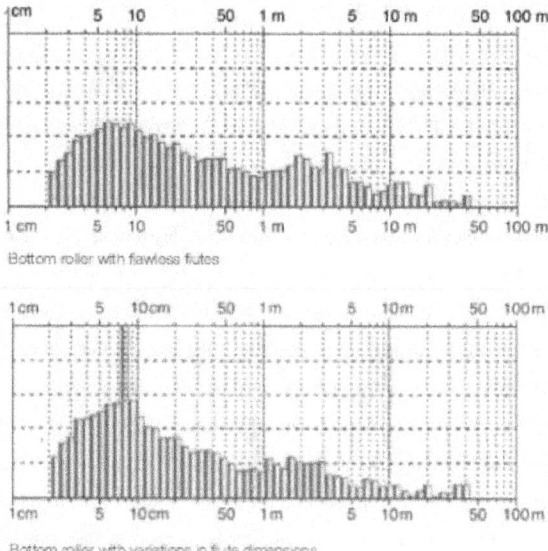

Bottom roller with flawless flutes

Bottom roller with variations in flute dimensions

The smooth running of the top roller with its direct contact to the roving influences the drafting result and therewith the yarn quality achieved.

6.3.2.2 Top Rollers:

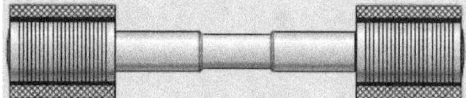

Top rollers are held strictly parallel and in perfect alignment with bottom rollers. Top rollers covered with rubber cots play a significant role in the control of drafting irregularities. The shore hardness of the rubber cots should be as per recommendations to control the fibers effectively. Shore hardness of top rollers depends upon the type of material and the process. Grinding of top roller cots should be performed with utmost care. Maintenance of top rollers such as top roller greasing and cots grinding should be done as per norms. The diameter should not be reduced by more than 3mm to ensure sufficient loading pressure. Top roller setting can be adjusted with the top arms in their loaded position. The fibre or dust accumulation in the top roller neck should be cleaned shift wise using picker gun. The dust accumulation in the neck of the top roller may resist the running of top roller which runs in contact with bottom roller. The usage of knife for clearing the roller lapping should be prohibited as it damages the cots surface.

6.3.2.3 Aprons, Cradle, Condensers and Spacer:

Efficient drafting requiring effective fibre speed control, especially that of the short fibres floating between the nips of the front and back rollers. Aprons are the one of the most effective means to control the floating fibres within the drafting zones. Apron wear is accelerated by high drafts and sliver linear density. It is essential that the aprons should extend as closely as possible to the nip line of the front rollers.

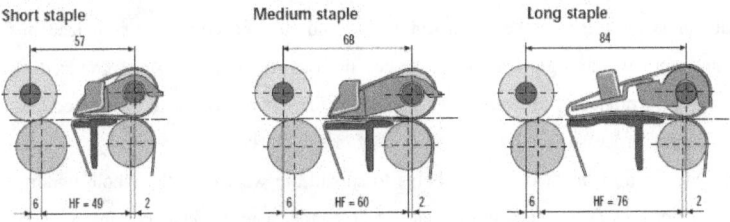

Cradles for Different Staple length

The top apron is short and made of synthetic rubber which has a thickness of about 1mm. Bottom apron is larger and made of same material as that of uppers. Basically, aprons made of synthetic rubber are made in endless tubular form whereas leather aprons are made in open strips and subsequently glued together to form an apron, the advantage of tabular construction is the lack of seam and uniform along its circumference. The top apron cradle ensures quick and trouble free replacement of the top aprons. The cradles can be easily fitted and removed. The top aprons are forced against the bottom aprons with the help of spring pressure. The combination of spring pressure and distance between top and bottom apron decides the intensity of fiber control.

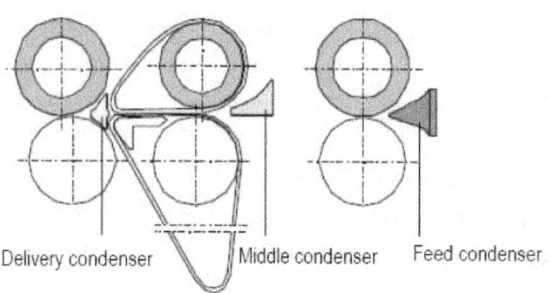

Delivery condenser   Middle condenser   Feed condenser

Condensers in Drafting Zone

Condensers placed in the drafting zone help to control the fibre strand from spreading apart during drafting. Condensers can be classified as inlet condenser and floating condenser. Inlet condenser is mounted on the reciprocating bar and floating condenser is placed in the main drafting zone which significantly influences the quality of the roving produced as well as running performance of the speed frame machine. The condenser size is selected based on the fibres being processed and the thickness of the material passing through the draft zone. Reciprocating movement of the bar helps to spread the wear over the whole width of the top roller rubber cots. The selection of smaller size condenser for a coarser hank material leads to uncontrolled stretching and fiber accumulations.

Standard condenser sizes and colours sliver feed weight				
	Ne 0.16 to 0.12		Ne 0.12 to 0.10	
Position	Opening	Colour	Opening	Colour
Feed	12 mm	Black	14 mm	Red
Middle	10 mm	Grey	12 mm	White
Delivery	10 mm	Beige	12 mm	White

Size of the Condenser

Larger condensers fails to control the fibres and deteriorates the roving evenness (CV %). The condensers with respect to its dimension are coded with different colours. A compact roving by use of front zone floating condenser at speed frame will bring down hairiness, as this will reduce strand width at ring frame. Floating condenser can be used behind front roller at speed frame without any working problems for finer hanks but with coarser hanks from short staple cottons choke up of condenser is encountered. Ishtiaque and Rengasamy reported that lower width of middle condenser improves most of the quality parameters and but decreases roving breakage rate. With a narrow condenser, fibres at the edge of the ribbon collide with the condenser edge and decelerate. The deceleration of these fibres and their resultant bucking disturb the movement of other fibres during drafting which deteriorates roving evenness. With a very wide condenser, the width of the fibre ribbon becomes too large and as a consequence inter-fiber frictional contacts decreases which lead to drafting irregularities. Yarn imperfections decreases with the decrease in condenser size. Ishtiaque and Rengasamy also reported that the yarn tenacity increases with the decrease in roving condenser size due to the compactness of the fibres.

The distance between top and bottom aprons is maintained by a small component called "cradle spacer" or "spacer" which is inserted between the nose bar of the bottom apron and the cradle edge of top apron. The selection of spacer for a process depends on the hank of the sliver, break draft and roving hank.

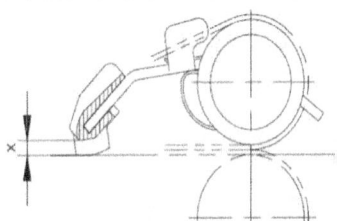

Cradle Spacers	X mm
green	2.5
pink	2.75
red	3
orange	3.25
brown	3.5
grey	4
yellow	5
blue	6
beige	7
black	8

Cradle Spacer – Colour and its dimensions

Roving Count Ne	Cradle Opening "X"
To 1.0	8.5 mm
1.1 to 1.8	6.5 mm
1.8 to 2.5	6.5 mm
Over 2.5	4.5 mm

Cradle Opening for Different Roving Counts

The reduction in size of the spacer may have several advantages and disadvantages as mentioned below:

Advantages	Disadvantages
• Improves the uniformity of roving • Reduces the imperfection level	• Affects the running behaviour of speed frame • Drafting problems • Generation of slubs due to over control of fibers

6.3.2.4 Top Arm Loading:

Spring loaded top arms are normally adopted in speed frame to get optimum pressure on top rollers enabling to achieve required quality and performance. The top arm pressure and roller setting influence the roving quality and subsequently yarn quality.

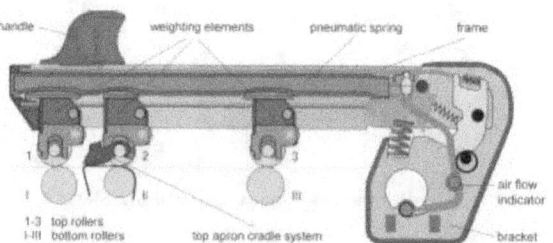

TEXparts® PK5025 Top arm for 3-roller drafting system

Yarn unevenness and imperfections shows an initial decrease up to a point with the increase of the above two parameters. In general, the moderate level of top arm pressure and roller setting gives better results. With the PK 5000 (TEXparts®) the weighting pressures on the top rollers are adjusted infinitely and centrally using a compressed air treatment system which provides constant loading at all spinning positions of the roving frame. The individual weighting arms are linked by connecting hoses to each other and to the air supply system. End pieces at both ends of the roving frame (first and last weighting arms) close off the air supply system. The pressure setting and system monitoring are performed centrally at the pneumatic unit installed in the machine control. The closed-circuit compressed air system of the PK 5000 ensures the same loading conditions in all arms and in all elements. The centralized pressure setting permits infinite and rapid adaptation to the technological requirements of the material to be spun.

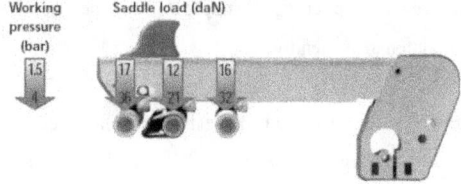

Top arm Loading

A study on influence of top roller loading conducted by Ishtiaque and Rengasamy reported that the increase in top roller loading initially decreases the roving U% and then increases. The reason attributed for this trend is the initial increase in top roller loading narrows down the gap between pressure fields of back and front beard of fibres and exert better control over the fibres which leads to reduction in U%. At higher top roller loading, there may be overlapping between the front and back pressure fields in the main drafting zone, which hinders the smooth fibre motion which results high roving U%. The above study also reported that the increase in top roller loading first reduces and then increases the roving breakage rate and imperfections. Top arm loading in speed frame do not have any significant influence on yarn tenacity as reported by this study.

6.3.3 Flyer and Spindle:

The rotation of flyer facilitates false twister fitted on its top to impart twist to the roving. A perfect balance during operation ensures consistent roving and prevents wear on parts. The flyers' special shape offers less air resistance, preventing roving breakages. Anti-static coating on flyer prevents fly accumulation, and light, precision-cast construction withstands the rigors of high-speed revolution. The features of a good flyer are summarized below:

- It should improve the quality of twisting by inserting false twist.
- No chance of occurrence of false draft by creating minimum resistance to flow of roving (i.e.) minimum surface frictional effect.
- The flyer should produce balance running condition especially at higher speed.
- The design and quality of metal of flyer should be such that there is no chance of spreading of flyer.
- There should be provision for slight changes in roving tension.
- There should be facility for easy and simple doffing operation.
- It should be maintenance friendly.

Presser arm is attached to the lower end of the flyer's hollow leg. The arm has to guide the roving from the exit of flyer leg to the package. The number of turns around the presser arm determines the roving tension and package hardness. If it is high then a compact package is obtained. The number of turns depends upon the fiber type and twist in the roving.

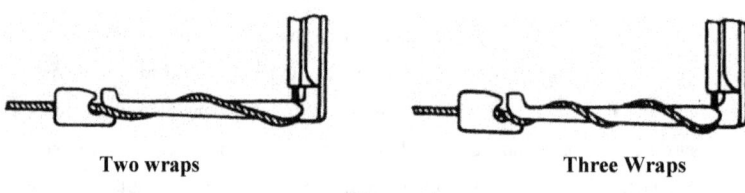

Two wraps   Three Wraps

Presser Arm

The spindle is a long vertical cylindrical shaft on which the flyer is mounted at the top. It is long steel shaft mounted at its lower end in a bearing and supported in the middle by a vertically reciprocating shaft of the package tube acting as the neck bearing. Flyer through spindle is driven by a set of gears housed in the spindle rail which is stationary. Spindle speed and flyer speed are same. The bobbin or the package over which the roving material is wound is loosely mounted over the spindle. The bobbin gets its drive from another set of gears housed in the bobbin rail which moves vertically up and down. The laying of roving coils onto the surface of the bobbin is achieved by the movement of bobbin rail. The flyer speeds employed normally range from about 1000 to 1400 rpm. According to the studies conducted by Shulz and Bohmer, roving irregularity increases with the increase in flyer speed due to higher flyer vibration. The reason attributed to the above trend is due to the fact that an increase in flyer vibration increases the centrifugal force and air resistance in the flyer leg and there will be a shearing action of roving with the flyer eye. Ishtiaque and Rengasamy reported that the increase in flyer speed increase the roving force in conjunction with the friction condition in the flyer leg which is responsible for more breakages.

6.4 Draft Distribution:

Drafting takes place by fibre straightening, fibre elongation and fibre sliding (relative movement). Draft given in the speed frame is usually calculated based on the roving hank produced. In a 3 over 3 drafting system, the first drafting zone, referred to as 'break-draft', is in the range of 1.03 to 2.03, while the main draft, is much higher, the total draft being the product of the two, generally ranging from about 4 to 20.

Recommended total draft range		
Fibre type	Preferred draft	Possible range
Short staple cotton	6 to 9	5 to 10
Medium staple	7 to 12	6 to 14
Long staple	9 to 18	8 to 18
Blends of cotton/ synthetics and man made fibres	7.5 to 12.5	7 to 13
100% synthetic fibres (polyester, viscose, acrylic, and nylon) up to 60 mm length	8 to 14	7.5 to 17

The break-draft facilitates the reduction of inter-fibre cohesion and frictional forces, thereby facilitating the fibres sliding past each other during the subsequent drafting. It is recommended to keep the break draft as low as possible. The problems associated with the higher break draft than recommended are given below:

- Requires higher drafting forces which can create vibrations in the back zone of the drafting system. Break draft may have to be as low as 1.022 to prevent roller vibrations.

- Tend to create roving irregularities such as thick and thin places.

Drafting generally introduces its own unevenness, increasing sliver unevenness to varying extents, good drafting requiring effective fibre speed control, particularly that of the short fibres floating between the nips of the front and back rollers. Aprons represent one of the most effective and popular means of controlling the movement of the floating fibres within the drafting zones. The process of attenuation of linear fibre assemblies by roller drafting causes a tension to be generated in the fibres in the drafting zone. Dutta reported that the drafting force and its variability are important characteristics that determine the irregularity added during drafting, the number of faults generated and the drafting failures. The force necessary to give rise to the average tension in the moving fibre mass in the drafting zone is referred to as the drafting force. Drafting force of roving has been found to affect spinning efficiency.

Fibre Parameters Influencing Drafting Force
• Fiber length
• Fiber fineness
• Fibre to fibre friction
• Fiber parallelization
• Packing factor
• Twist
• Fiber irregularity

Machine Parameters Influencing Drafting Force
• Draft ratio
• Drafting speed
• Roller Setting

According to the investigation by Das & Ishtiaque, the drafting force initially increases with the draft and then declines sharply as the draft increases further. This is due to the fact that at the lower level of draft, very little fibre slippage occurs due to elastic behaviour of fibre strand and the fibres are simply straightening out (removal of crimp and hooks). The maximum drafting force is observed at different draft for different roller settings. With further increase in draft, the principle modes of roving deformation is due to the sliding of fibres relative to one another because the static friction is fully overcomed and hence after the peak region the drafting force declines quickly. At the higher level of draft, the drafting force generated is due to the dynamic friction of fibre which is lesser than the static friction. At higher draft and drafting speed, the control over the fibres goes down and chances of fibre shuffling become less, thereby reducing the drafting force. Das and Ishtiaque reported that the drafting force always reduces with the increase in roller setting. The above trend is due to the fact that at lower roller setting, the control over the movement of floating fibres becomes more because of the more inter-fibre cohesion. But as the roller setting increases the inter-fibre cohesion goes down and hence drafting force reduces.

6.5 Twist:

Twisting in the speed frame is the process of rotating the fibrous strand about its own axis so that the fibers are arranged in a spiral form and thus bind each other together. The purpose of providing twist in roving is to give the strand sufficient strength to withstand the strain during unwinding in the creel of the ring frame. The insertion of twist is achieved by the rotation of the flyer. Twist level depends on flyer speed and delivery speed of the speed frame. The increase in twist reduces the productive capacity of the machine, so it is generally used in as limited range as possible. The relationship between the twist and above factors is given below:

Twist = Flyer speed or Spindle Speed (rpm) / Delivery speed (meters per min)

False twisting devices are used on the flyers to add false twist when the roving is being twisted between the front roller and the flyer. Because of this supplementary twist, the roving is strongly twisted and this reduces the breakage rate. False twisting device is also called as "twist crown"

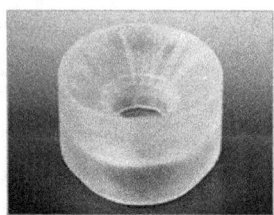

The level of twist imparted in speed frame process varies with the staple of the cotton and the hank of the roving. Longer cotton requires less twist because individual fibers extend further in the strand and so help to bind them together more securely than do short fibres. Finer roving require more twist compared to coarser rovings. In mill practice, the level of twist is normally judged by the way the roving acts. Roving must have sufficient twist to give strength to turn the bobbin in the creel of the ring frame without having it break. The lower twist level reduces the production and efficiency of the machine due to higher roving breakages. Higher twist in speed frame reduces the production rate and so increases the cost of production. Ishtiaque and Atal Vijay reported that the increase in roving twist increases the inter-fibre friction due to more contact area, which creates problem during ring frame drafting and ultimately deteriorates the yarn quality. Das have concluded that the yarn irregularity and total imperfections increase with the increase in fibre-to-fibre friction. In another study, Das have shown that the increase in bottom apron slippage as a result of increase in roving TM. Balasubramanian states that one of the reasons for stretch of strand in the creel is low roving twist. Edger Bay and Frank baier reported that high twist roving can be withdrawn without creel stretch in ring frame. The high twist roving have the advantage of large bobbin sizes and fewer roving brakes in speed frame so that roving uniformity can be improved. Ching Iuan Su and Kuo Jung Lo discussed that the high twist roving could not be sufficiently drafted in the back roller zone, resulting in thick places in the spun yarn. Basu reported that with the increase in break draft, the yarn imperfection also increases and they further state the yarn imperfection increases due to the increasing roving twist multiplier for same yarn count. The problems associated with the high roving twist w.r.t process are:

- Roving may not get properly drafted in ring frame break draft zone.
- Undrafted roving will be pulled through the system and called as "Hard Ends"
- More ends down due to hard ends.

- Ends down at the roving frame

 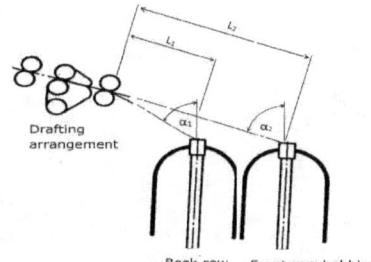

Front and back row of Flyer

The speed frames are generally fitted with two rows of the flyer. The amount of false twist inserted by the flyer inlet is dependent on the contact angle of the roving length between the flyer and the front drafting rollers; the smaller the angle, the higher the false twist. The back row flyers are nearer to the drafting rollers and has the greater contact angle. The false twist given by the back row flyer does not provide sufficient cohesion to the roving which reduces the roving hank slightly. There is a significant count variation between the rovings produced from the front row and back row flyers. In order to overcome this problem, modern speed frames have the back row flyer fitted with a raised false twister attachment providing the contact angles of the two row to almost similar.

6.6 Bobbin Formation:

Bobbin formation is the one of important task performed in the speed frame machine. The roving is drawn from the front roller through the flyer and onto the bobbin. The bobbin has a higher surface speed than the flyer which winds the twisted roving onto the bobbin. The centrifugal forces increase as the bobbin diameter increases. The rate of winding compared with the front roller delivery rate controls the roving tension. There is no precise method for measuring roving tension.

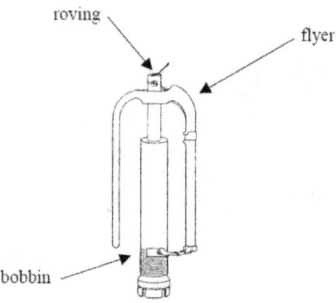

Flyer and Bobbin

It is customary to say in spinning mills that tension is "tight" if there is considerable pull between the front roller and the bobbin. On the other hand the tension is "slack" or "loose" when the bobbin hardly winds the roving delivered by front roller. The setting of proper roving tension is crucial in the bobbin formation depends upon mostly by judgment and experience.

6.6.1 Taper Formation:

The bobbin obtained from speed frame is a cylindrical package with conical ends. The drafted fibrous strand is wound onto an empty package which is usually made of plastic. Each roving layer is made of helical coils arranged close to each other which are attained by the up and down movement of bobbin. After the completion of one layer, the bobbin's direction of movement is reversed to start a next layer. The up and down movement of the bobbin for layering a layer is called as "traverse". The taper is achieved by reducing the amount of traverse after completion of each layer.

We know, coils per minute = Bobbin speed − Spindle speed

$$\text{Therefore, coils per inch} = \frac{\text{Coils per minute}}{\text{Rate of lift in inches per minute}}$$

Again, rate of lift = RPM of lifter shaft wheel x Teeth of lifter wheel x pitch of the lifter wheel

$$\text{Rpm of lifter shaft wheel} = \text{Motor rpm} \times \frac{\text{Driver}}{\text{Driven}}$$

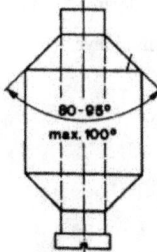 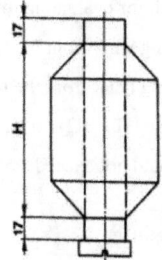

6.7 Quality Control of Roving:

The control of roving quality is very essential since speed frame is the final stage of spinning preparatory process. Most of the sources of yarn faults are mainly due to the bad roving bobbin quality. The higher irregularity of roving tends to severely affect the tensile properties of the yarn produced from it. The various quality aspects of roving are discussed in detailed manner in the below sections:

6.7.1 Ratching:

The main source of long term variation in the speed frame is ratching. The tension draft given in speed frame at the commencement of doff should be less than 1%. The initial layers of bobbin are normally prone to degree of ratching. The bobbin rail should be set up or down to wind with a full layer to overcome the occurrence of ratching. Roving tension at the start of doff depends upon bare bobbin diameter and cone drum belt position. A 5% ratching in roving tend to increase 15% variations in the yarn. Ratching can be determined by comparing the count of roving during start and end of doff. The position of the cone-drum belt at the time of full doff also gives an indication of ratching. If the belt is near its extreme position, the ratchet wheel is almost correct.

6.7.1.1 Procedure to determine ratching% in roving:

- Collect the full bobbins from the spindles during doff in the speed frame.
- Run the machine for 2-5 minutes.
- Collect the bobbins with 2 initial layers from the same spindles immediately.

- Sample length for evaluating roving hank is 15-30 yards
- Find the roving hank from the bobbins with 2 initial layers(H1)
- Find the roving hank from the full bobbin (H2)
- Ratching% = ((H1 − H2) / ((H1 + H2)/2)) x 100
- Ratching should not exceed 1.5%

6.7.2 Roving Strength:

Roving strength is an important quality parameter which is required for trouble free unwinding in ring frame. It plays a significant role in the optimization of roving twist multiplier. Some of the poor work practices attributed for poor roving strength are:

- Improper sliver piecing at creel
- Improper roving piecing
- Stretch at creel
- Poor condition of spindle

6.7.3 Count C.V%:

Roving hank or count C.V% can be categorized as within-bobbin C.V% and between-bobbin C.V%. The sample length taken for testing roving count C.V% is normally 15 yards. The count C.V% of roving falls in the range of 1.5% to 2.0 % under good working conditions. Ratching within lengths of about 100 to 300m of rove would produce a higher effect on within-bobbin C.V%, while ratching over higher lengths would affect between bobbin C.V% to a great extent. Tension variations from start to end of the doff causes roving hank variation. The various reasons for the occurrence of the count variation within- and between-bobbin are summarized as below:

Various Reasons for Count C.V%	
Within-bobbin	**Between-bobbin**
• Uneven tension during bobbin built-up • Sliver splitting and stretching at creel zone. • Roller lapping • Defective Spindle • High C.V% in sliver	• Variation in drafting pressure • Quality of sliver (1m CV %)

6.7.4 Unevenness:

Unevenness or irregularity of roving has predominant influence on the yarn quality. The rovings with high irregularity cause more erratic movement of fibres resulting in more drafting force variability. The control of short term irregularity and end breaks is very important in speed frame. The various factors influencing the unevenness of the roving are:

- Quality of cots,
- Aprons
- Spacer size
- Draft
- Condenser size
- Break draft,
- Top arm pressure

6.8 Defects in Roving:

6.8.1 Higher U% of rove:

Causes	Remedies
• Inadequate top arm pressures • Improper settings • Worn out gears or bearings • Grooved top rollers, tilted top rollers • Wrong selection of condensers • Worn out aprons • Poor cleaning of draft zone • Higher stretch • Uneven feed material	Refer the spectrogram and check the spindle and the feed material before taking any action. If any machine parts found faulty, correct it.

• Sliver splitting in creel, jerks in creel movement • Vibrations in the machines	

6.8.2 Higher roving breakages:

Causes	Remedies
• Uneven sliver	• Supply high quality sliver with lower C.V%
• Worn out parts, damaged machine parts & vibrations	• Replace worn out parts; proper maintenance of machine parts
• Insufficient twist	• Impart twist as per norms
• Improper draft distribution	• Proper draft distribution
• Fluctuations in R.H%	• Proper follow up of R.H%
• Rough surface in the flyer tube	• Polish the surface of flyer tube
• Improper build of bobbins	• Proper maintenance of builder motion
• Improper piecing of sliver	• Proper work practices for correct piecing
• Uncontrolled air current, etc.	• Arrest uncontrolled air currents
• Higher spindle speed than recommended	• Speeds and settings as per recommendations

6.8.3 Soft bobbins:

Causes	Remedies
• Too finer hank	• Roving hank should be optimized.
• Singles or a finer drawing hank	• Avoid singles; Optimize drawing hank
• Less number of turns on presser arm	• Number of turns on presser arm as per recommendations
• Belt shift on cone drum faster than required	• Optimize the rate of belt shift on cone drum
• Lower TPI	• Optimum TPI
• Lower relative humidity	• Maintain proper R.H%

6.8.4 Lashing-in:

Causes	Remedies
• Broken end joins to adjacent end and creates lashing-in	• Provision of separators, roving end catcher solve this problem • Set the suction tube near to the front roller nip • Reduce the end breakage rate

6.8.5 Hard bobbins:

Causes	Remedies
• Coarser hank • Doubles or coarser draw frame hank • Lower top arm loading • Higher twist • Lesser movement of belt on cone drums • More number of turns on flyer presser arm • Higher R.H%	• Optimize the roving hank • Correct doubles and maintain proper draw frame hank • Optimize twist • Proper rate of movement of belt on cone drums • Recommended number of turns on flyer presser arm • Maintain proper R.H%

6.8.4 Oozed out bobbins:

Causes	Remedies
• Malfunction of reversing bevels in the builder motion • Stopping the machine when the bobbin rails are in extreme positions, and • Jumping bobbin	• Correct the malfunction in builder motion • Stop the machine when bobbin rails are in middle position • Proper quality of empty bobbin to be maintained.

6.8.5 High roving Count C.V%:

Causes	Remedies
• Insufficient top arm pressure • Too high a break draft • Improper selection of condenser guides • Vibration of roving bobbin while running • Sliver stretch at creel • High variation in empty bobbin diameter • Wrong selection of winding-on wheel and ratchet wheel • Improper shifting of cone drum belt • Disturbance in the movement of aprons • Rough spots on flyer • Bobbin rail movement is not uniform	• Set top arm pressure, break draft and condenser guides as per norms • Check the proper seating of bobbin on pin. • Control stretch at creel • Proper selection of wheels • Check the belt position on cone drum and correct. • Proper cleaning of drafting zone • Proper maintenance of flyer and bobbin rail.

6.8.6 Roller lapping:

Causes	Remedies
• Too high spindle speed and draft • Cuts or damages in the top roller cots • Damages in apron, condenser, bottom roller • Wrong choice of spacer • Wider setting at the back zone • Improper R.H%	• Set speeds, settings and drafts as per norms • Avoid usage of knife while clearing roller lapping. • Proper maintenance of top roller cots, apron, condenser, bottom roller • Select spacer size as per norms • Maintain R.H% as per norms

6.8.7 Slubs:

Causes	Remedies
• Too high end breakages • Waste accumulation at creel, drafting zone and flyer • Wrong selection of spacer • Too low break draft • Closer setting at the back zone • Top and bottom clearer not functioning properly	• Minimize end breakages • Proper cleaning of machine parts during shifts • Proper selection of spacer, break draft and back zone setting. • Ensure proper running of clearer roller

6.9 Technological developments in Speed frame:

The machinery developments in speed frame are significantly low compared to other machines of the spinning process. This is proved by the fact that the spindle speed of speed frame has attained only 1500 rpm as on date compared to 600 rpm in 1950s. In terms of production and quality, increase in the roving bobbin diameter from 4" to 7" and lift from 8" to 16", use of straight cone drum instead of hyperbolic cone drum for better control over the roving tension etc... are the significant developments occurred in the last two decades. Almost all the latest speed frames are fitted with closed (AC Type) flyers that are used to overcome the problem of air drag on roving. These flyers are aerodynamically balanced and are of light weight. The roving frames are equipped with auto doffing system that, apart from avoiding man handling, reduces doffing time. Even the Toyota claims to have auto doffing on FL100 roving frame in record time of 3½ minutes. Roving bobbins auto doffing and transportation, to the ring spinning through over head rails, becomes a standard feature of the roving frame.

In multi-motor drive system, drafting rollers, flyers, bobbins and bobbin rail are driven directly by individual servomotors and are synchronized throughout package build by the control system. The advantages of this system are like, no need of heavy counter weight for bobbin rail balancing and differential gear, reduced maintenance, lower energy consumption, etc.

Roving Tension sensor (Rieter F15/F35 roving frame)

Roving tension sensors measure and control the roving tension (constant) throughout the bobbin build. These tension sensors do not actually contact the roving while measuring the tension. The tension is measured at periodic intervals and the required change in tension is actuated by changing the bobbin speed through servomotor. Rieter F15/F35 roving frame, Zinser 668 roving frame, Marzoli FTN roving frame, Lakshmi LFS 1660 speed frame and Toyota FL100 roving frame have incorporated roving tension sensor on their machines.

 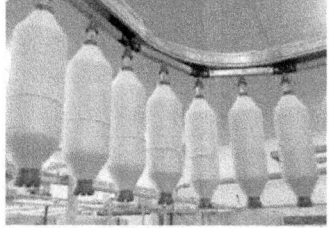

Roving Transport System

Automated bobbin transport offers the advantages of labor savings and a substantial increase in bobbin quality. The roving bobbin is one of the most delicate intermediate products to handle for two reasons: the roving wound around the bobbin is completely unprotected and is therefore highly susceptible to damage; and all roving defects are transferred to the yarn and cannot be corrected. Automated bobbin transport eliminates the need both to handle the bobbin or touch the textile product and to maintain intermediate storage areas, where bobbins can accidentally age, get dirty and deteriorate. The train of bobbins is automatically transported from speed frame to the storage area and respective ring frame by selecting the appropriate program in the PLC. Empty bobbins in the ring frame are manually interchanged with full bobbins from BTS by the operator and empty bobbins are transported back to speed frame automatically.

References:

1. Peter.R.Lord, Hand book of yarn technology: Technology, Science and Economics, Wood head Publishing Ltd and CRC Press LLC, 2003.
2. Carl.A.Lawrence, Fundamentals of Spun yarn technology, CRC press, 2003.
3. S.Gordon, Y.L.Hsieh, Cotton: Science and Technology, Wood head publications & CRC press, 2007.
4. Klein, W. A Practical Guide to Opening and Carding, Manual of Textile Technology-2, Textile Institute, 1987.
5. Grosberg, P. and Iype, C., Yarn Production: Theoretical Aspects, The Textile Institute, Manchester UK, 1999.
6. Oxtoby, E., Spun Yarn Technology, Butterworths & Co., London, 1987.
7. Artzt, P., Short Staple Spinning: Quality Assurance and Increased Productivity, ITB International Textile Bulletin, 2003, 49 (6), 10.
8. Cotton yarn Manufacturing, www.ilo.org/oshenc/part-xiv/.../880-cotton-yarn-manufacturing
9. Rakesh S Chauhan, Yarn hairiness: Measurement, effect & consequences, Indian Textile Journal, February, 2009.
10. PK5000 Innovative drafting system technology, Texparts Information brochure.
11. Novibra: Bottom Rollers, www.novibra.com/index.php?id=119
12. Drafting aprons – Torque value – Ring frame - Cotton Yarn Market, www.cottonyarnmarket.net/OASMTP/Drafting%20aprons.pdf
13. H. Schwippl, Technology Handbook for Service Engineers, Rieter, Edition 2006.

14. Saiyed M. Ishtiaque, Apurba Das, Ritesh Niyogi, Optimization of Fiber Friction, Top Arm Pressure and Roller Setting at Various Drafting Stages, Textile Research Journal December 2006 vol. 76 no.12 913-921.

15. Shulz G., Melliand Textilber, 68(8), 1987, E238.

16. Bohmer I., Melliand Textilber, 73(6), 1992, E242.

17. A Das, S M Ishtiaque & Rajesh Kumar, Study on drafting force of roving: Part I – Effect of process variables, Indian Journal of Fibre & Textile Research, Vol.29-June 2004-pp 173-17

18. Dutta B, Salhotra K R & Qureshi A W, Blended textiles, paper presented at the 38th All India Textile Conference, Mumbai, November 1981.

19. Das A, Ishtiaque S M, Yadav P & Kumar Rajesh, Design and development of draftometer and a critical study on drafting force of roving, paper presented at the 44th Joint Technological Conference of ATIRA, BTRA, SITRA and NITRA, Coimbatore, March 2003.

20. Graham J S & Bragg C K, Text Res J, 42 (1972) 180.

21. Plonsker H R & Backer S, Text Res J, 37 (1967) 673.

22. Ishtiaque S M and Atal Vijay (1994): Optimisation of Ring Frame Parameters for Coarser Preparatory, Indian Journal of Fibre & Textile Research, Vol 19, No: 4, pp 239 - 246.

23. Deluca B and Thibodeaux D P (1992): The Relative Importance of Fibre Friction and Torsional and Bending Rigidities in Cotton Sliver, Roving and Yarn, Textile Research Journal, Vol 62, No: 4, pp 192 - 196.

24. Das A, Yadav P and Ishtiaque S M (2002): Apron Slippage in Ring Frame: Part II - Factors Affecting Apron Slippage and Their Effect on Yarn Quality, Indian Journal of Fibre & Textile Research, Vol 27, pp 135 - 141.

25. Balasubramanian N (2006): Controlling Count Variation in Yarn, http://Business.vsnl.com/ balasubramanian/index.html.

26. Edgar Bay and Frank Baier (1996): Modern Drafting System for Improving Flexibility in Ring Spinning, International Textile Bulletin, Vol.42, pp 64 - 70.

27. Ching Iuan Su and Kuo Jung Lo (2000): Optimum Drafting Conditions of Fine - Denier Polyester Spun Yarn, Text Res J, Vol 70, No: 2, pp 93 - 97.

28. Arindam Basu and Rajanna Gotipamul (2005): Effect of Some Ring Spinning and Winding Parameters on Extra-sensitive Yarn Imperfections, Indian Journal of Fibre & Textile Research, Vol.30, No.2, pp 211 - 214.

29. A Peer Mohamed and D Veerasubramanian, Roving twist & its significance, The Indian Textile Journal, June, 2009.

30. Reeti Pal Singh and V K Kothari, Developments in comber, speed frame & ring frame, The Indian Textile Journal, May , 2009.

31. Zinser 670 RoveMat: Roving Frame with Integrated Doffer, Zinser Textilmaschinen GmbH, Information brochure.

32. Roving Frame F 15/F 35: The Solution for Your Spinning Preparation, Rieter Textile System, Information brochure.

33. Zinser 668: The Roving Frame Solution for Simplified Setting and More Flexibility, Zinser Textilmaschinen GmbH, Information brochure.

34. Spinning Section: Roving Frame FTN, Marzoli Spa, Italy, Information brochure.

35. Spinning Value: Speed Frame LFS 1660, Lakshmi Machine Works Ltd, Information brochure.

36. Toyota Roving Frame FL100, Toyota Textile Machinery Division, Information brochure.

37. Piergiuseppe Bullio, Janet Bealer Rodie, A Question Of Change, Textile World, November-December, 2006.

38. B.Purushothama, Training and Development of Technical Staff in the Textile Industry, Wood head Publishing India Pvt. Ltd., New Delhi, 2012.

39. B.Purushothama, A Practical Guide for Quality Management in Spinning, Wood head Publishing, India, 2012.

40. Ratnam, T. V and Chellamani, K. P., Quality Control in Spinning, The South India Textile Research Association, Coimbatore, Third Revised ed. 1999.

41. Shankaranarayana K.S., Proceedings of 14th Joint Technological Conference (ATIRA/BTRA/SITRA), 1973, p.55.

42. P.R.Lord and G.Grover, Roller Drafting, Textile Progress, Volume 23, No.4, Textile Institute, 1993.

www.ingramcontent.com/pod-product-compliance
Lightning Source LLC
Chambersburg PA
CBHW070304190526
45169CB00004B/1522